Introduction to Quantum Computing

Published 2023 by River Publishers

River Publishers

Alsbjergvej 10, 9260 Gistrup, Denmark

www.riverpublishers.com

Distributed exclusively by Routledge

605 Third Avenue, New York, NY 10017, USA

4 Park Square, Milton Park, Abingdon, Oxon OX14 4RN

Introduction to Quantum Computing / by Ahmed Banafa.

Routledge is an imprint of the Taylor & Francis Group, an informa business

ISBN 978-87-7022-841-1 (paperback)

ISBN 978-10-0096-238-3 (online)

ISBN 978-1-003-44023-9 (ebook master)

A Publication in the River Publishers series
RAPIDS SERIES IN COMPUTING AND INFORMATION SCIENCE AND TECHNOLOGY

Introduction to Quantum Computing

Ahmed Banafa

San Jose State University, USA

River Publishers

Routledge
Taylor & Francis Group
NEW YORK AND LONDON

In the loving memory of my son Malik

Contents

Preface

Quantum Computing is the area of study focused on developing computer technology based on the principles of quantum theory. Tens of billions of public and private capitals are being invested in quantum technologies. Countries across the world have realized that quantum technologies can be a major disruptor of existing businesses, they have collectively invested billions of dollars in quantum research and applications. In this book you will learn the difference between quantum computing and classic computing, also different categories of quantum computing will be discussed in details, applications of quantum computing in AI, IoT, Blockchain, communications, and encryption will be covered, in addition, quantum internet, quantum cryptography, quantum teleportation, and post-quantum technologies will be explained.

This is a list of the chapters of the book:

Chapter 1 : What is Quantum Computing?
Chapter 2: Quantum Cryptography
Chapter 3: Quantum Internet
Chapter 4 : Quantum Teleportation
Chapter 5: Quantum Computing and the IoT
Chapter 6: Quantum Computing and Blockchain: Myths and Facts
Chapter 7 : Quantum Computing and AI: A Mega-Buzzword
Chapter 8: Quantum Computing Trends

Audience

This is book is for everyone who would like to have a good understanding of Quantum Computing and its applications and its relationship with business operations, and also gain insight to other transformative technologies like IoT, cloud computing, deep learning, Blockchain, Big Data and wearable technologies. The audience includes: C-Suite executives, IT managers, marketing and sales professionals, lawyers, product and project managers, business professionals, journalists, students.

Acknowledgment

I am grateful for all the support I received from my family during the stages of writing this book.

About the Author

Professor Ahmed Banafa has extensive experience in research, operations and management, with a focus on IoT, Blockchain, Cybersecurity and AI. He is the recipient of the Certificate of Honor from the City and County of San Francisco and Author & Artist Award 2019 of San Jose State University. He was named as No. 1 tech voice to follow, technology fortune teller and influencer by LinkedIn in 2018, his research has featured on Forbes, IEEE and MIT Technology Review, and he has been interviewed by ABC, CBS, NBC, CNN, BBC, NPR, Washington Post, and Fox. He is a member of the MIT Technology Review Global Panel. He is the author of the book *Secure and Smart Internet of Things (IoT) using Blockchain and Artificial Intelligence (AI)* which won three awards – the San Jose State University Author and Artist Award, One of the Best Technology Books of all Time Award, and One of the Best AI Models Books of All Time Award. His second book was *Blockchain Technology and Applications*, which won San Jose State University Author and Artist, One of the Best New Private Blockchain Books and is used at Stanford University and other prestigious schools in the USA. His most recent book is *Quantum Computing*. He studied Electrical Engineering at Lehigh University, Cybersecurity at Harvard University and Digital Transformation at Massachusetts Institute of Technology (MIT).

1

What is Quantum Computing?

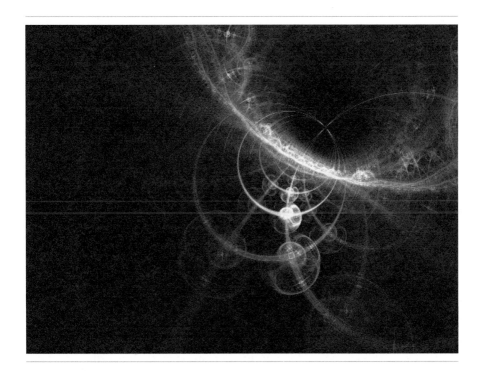

Quantum computing is the area of study focused on developing computer technology based on the principles of quantum theory. The quantum computer, following the laws of quantum physics, would gain enormous processing power through the ability to be in multiple states and to perform tasks using all possible permutations simultaneously.

1.1 A Comparison of Classical and Quantum Computing

Classical computing relies, at its ultimate level, on principles expressed by Boolean algebra. Data must be processed in an exclusive binary state at any point in time or bits. While the time that each transistor or capacitor needs to be either in 0 or 1 before switching states is now measurable in billionths of a second, there is still a limit as to how quickly these devices can be made to switch state. As we progress to smaller and faster circuits, we begin to reach the physical limits of materials and the threshold for classical laws of physics to apply. Beyond this, the quantum world takes over. [1]

In a quantum computer, a number of elemental particles such as electrons or photons can be used with either their *charge* or *polarization* acting as a representation of 0 and/or 1. Each of these particles is known as a quantum bit, or *qubit*; the nature and behavior of these particles form the basis of quantum computing.

1.2 Quantum Superposition and Entanglement

The two most relevant aspects of quantum physics are the principles of *superposition* and *entanglement*.

Superposition: Think of a qubit as an electron in a magnetic field. The electron's spin may be either in alignment with the field, which is known as a spin-up state, or opposite to the field, which is known as a spin-down state. According to quantum law, the particle enters a superposition of states, in which it behaves as if it were in both states simultaneously. Each qubit utilized could take a superposition of both 0 and 1.

Entanglement: Particles that have interacted at some point retain a type of connection and can be entangled with each other in pairs, in a process known as *correlation*. Knowing the spin state of one entangled particle – up or down – allows one to know that the spin of its mate is in the opposite direction. Quantum entanglement allows qubits that are separated by incredible distances to interact with each other instantaneously (not limited to the speed of light). No matter how great the distance between the correlated particles, they will remain entangled as long as they are isolated.

Taken together, quantum superposition and entanglement create an enormously enhanced computing power. Where a 2-bit register in an ordinary computer can store only one of four binary configurations (00, 01, 10, or 11) at any given time, a 2-qubit register in a quantum computer can store all four numbers simultaneously, because each qubit represents two values. If more qubits are added, the increased capacity is expanded exponentially. [2]

1.3 Difficulties with Quantum Computers

- **Interference:** During the computation phase of a quantum calculation, the slightest disturbance in a quantum system (say a stray photon or a wave of EM radiation) causes the quantum computation to collapse, a process known as de-coherence. A quantum computer must be totally isolated from all external interference during the computation phase.
- **Error correction:** Given the nature of quantum computing, error correction is ultra critical – even a single error in a calculation can cause the validity of the entire computation to collapse.
- **Output observance:** Closely related to the above two, retrieving output data after a quantum calculation is complete risks corrupting the data. [3]

1.4 The Future of Quantum Computing

The biggest and most important one is the ability to factorize a very large number into two prime numbers. This is really important because this is what almost all *encryption* of internet applications uses and can be de-encrypted. A quantum computer should be able to calculate the positions of individual atoms in very large molecules like polymers and in viruses relatively quickly. If you have a quantum computer you could use it, and the way that the particles interact with each other, to develop drugs and understand how molecules work a bit better. Even though there are many problems to overcome, the breakthroughs in the last 15 years, and especially in the last 3, have made some form of practical quantum computing possible. However, the potential that this technology offers is attracting tremendous interest from both the government and the private sector. It is this potential that is rapidly breaking down the barriers to this technology, but whether all barriers can be broken, and when, is very much an open question. [4, 5]

2

Quantum Cryptography

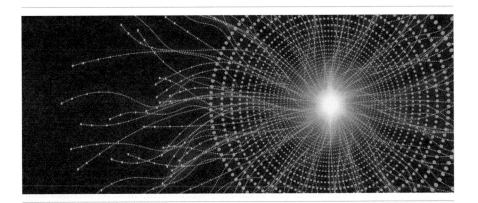

Quantum cryptography uses physics to develop a cryptosystem completely secure against being compromised, without the sender or the receiver of messages being known. The word *quantum* itself refers to the most fundamental behavior of the smallest particles of matter and energy.

Quantum cryptography is different from traditional cryptographic systems in that it relies more on physics, rather than mathematics, as a key aspect of its security model. [7]

Essentially, quantum cryptography is based on the usage of individual particles/waves of light (photon) and their intrinsic quantum properties to develop an unbreakable cryptosystem (because it is impossible to measure the quantum state of any system without disturbing that system).

Quantum cryptography uses photons to transmit a key. Once the key is transmitted, coding and encoding using the normal secret-key method can take place. But how does a photon become a key? How do you attach information to a photon's spin? [8]

This is where *binary code* comes into play. Each type of a photon's spin represents one piece of information – usually a 1 or a 0, for binary code. This code uses strings of 1s and 0s to create a coherent message. For example, 11100100110 could correspond to h-e-l-l-o. So a binary code can be assigned to each photon – for example, a photon that has a *vertical spin* (|) can be assigned a 1.

"If you build it correctly, no hacker can hack the system. The question is what it means to build it correctly," said physicist Renato Renner from the Institute of Theoretical Physics in Zurich. [9]

Regular, non-quantum encryption can work in a variety of ways but generally a message is scrambled and can only be unscrambled using a secret key. The trick is to make sure that whomever you're trying to hide your communication from doesn't get their hands on your secret key. Cracking the private key in a modern crypto system would generally require figuring out the factors of a number that is the product of two insanely huge prime numbers. The numbers are chosen to be so large that, with the given processing power of computers, it would take longer than the lifetime of the universe for an algorithm to factor their product.

However, such encryption techniques have their vulnerabilities. Certain products – called weak keys – happen to be easier to factor than others. Also, Moore's Law continually ups the processing power of our computers. Even more importantly, mathematicians are constantly developing new algorithms that allow for easier factorization. Quantum cryptography avoids all these issues. Here, the key is encrypted into a series of photons that get passed between two parties trying to share secret information. The Heisenberg uncertainty principle dictates that an adversary can't look at these photons without changing or destroying them.

"In this case, it doesn't matter what technology the adversary has, they'll never be able to break the laws of physics," said physicist Richard Hughes of Los Alamos National Laboratory in New Mexico, who works on quantum cryptography. [10]

2.1 Problems with Using Quantum Cryptography

However, in practice, quantum cryptography comes with its own load of weaknesses. It was recognized in 2010, for instance, that a hacker could blind a detector with a strong pulse, rendering it unable to see the secret-keeping photons.

Renner points to many other problems. Photons are often generated using a laser tuned to such a low intensity that it is producing one single photon at a time. There is a certain probability that the laser will make a photon encoded with your secret information and then a second photon with that same information. In this case, all an enemy has to do is steal that second photon and they could gain access to your data while you would be none the wiser. Alternatively, noticing when a single photon has arrived can be tricky. Detectors might not register that a particle has hit them, making you think that your system has been hacked when it is really quite secure. [11]

"If we had better control over quantum systems than we have with today's technology" then perhaps quantum cryptography could be less susceptible to problems," said Renner. But such advances are at least 10 years away. Still, he added, no system is 100% perfect and even more advanced technology will always deviate from theory in some ways. A clever hacker will always find a way to exploit such security holes.

Any encryption method will only be *as secure as the humans running it*, added Hughes. Whenever someone claims that a particular technology "is fundamentally unbreakable, people will say that's snake oil," he said. "Nothing is unbreakable." [12]

3

Quantum Internet

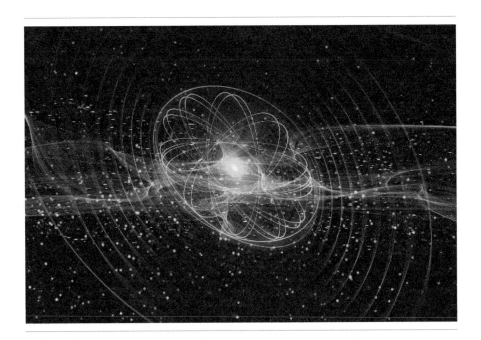

Building a quantum internet is a key ambition for many countries around the world as such a breakthrough will give them competitive advantage in a promising disruptive technology, and opens a new world of innovations and unlimited possibilities.

Recently the US Department of Energy (DoE) published the first blueprint of its kind, laying out a step-by-step strategy to make the quantum internet dream come true. The main goal is to make it impervious to any cyber hacking. It will "metamorphosize our entire way of life," says the Department of Energy. Nearly $625 million in federal funding is expected to be allocated to the project. A quantum internet would be able to transmit large volumes of data across immense distances at a rate that exceeds the *speed of light*. You can imagine all the applications that could benefit from such speed.

Traditional computer data is coded in either zeros or ones. Quantum information is superimposed in both zeros and ones simultaneously. Academics, researchers, and IT professionals will need to create devices for the infrastructure of quantum internet including: quantum routers, repeaters, gateways, hubs, and other quantum tools. A whole new industry will be born based on the idea of a quantum internet existing in parallel with the current ecosystem of companies we have in a regular internet. The "traditional internet", as the regular internet is sometimes called, will still exist. It is expected that large organizations will rely on the quantum internet to safeguard data, but that individual consumers will continue to use the classical internet. [13]

Experts predict that the financial sector will benefit from the quantum internet when it comes to securing online transactions. The healthcare sectors and the public sectors are also expected to see benefits. In addition to providing a faster, safer internet experience, quantum computing will better position organizations to solve complex problems, like supply chain management. Furthermore, it will expedite the exchange of vast amounts of data, and carrying out large-scale sensing experiments in astronomy, materials discovery, and life sciences. [13, 15]

But first let's explain some of the basic terms of the quantum world: *Quantum computing* is the area of study focused on developing computer technology based on the principles of *quantum theory*. The quantum computer, following the laws of quantum physics, would gain *enormous processing power* through the ability to be in multiple states, and to perform tasks using all possible permutations simultaneously. [14]

In a quantum computer, a number of elemental particles such as *electrons or photons* can be used with either their *charge* or *polarization* acting as a representation of 0 and/or 1. Each of these particles is known as a *quantum bit*, or *qubit*, the nature and behavior of these particles form the basis of quantum computing. [14]

3.1 What is the Quantum Internet?

The quantum internet is a network that will let quantum devices exchange some information within an environment that harnesses the odd laws of quantum mechanics. In theory, this would lend the quantum internet unprecedented capabilities that are impossible to carry out with today's web applications.

In the quantum world, data can be encoded in the state of qubits, which can be created in quantum devices like a quantum computer or a quantum processor. And the quantum internet, in simple terms, will involve sending qubits across a network of multiple quantum devices that are physically separated. Crucially, all of this would happen thanks to the wild properties that are unique to quantum states.

That might sound similar to the standard internet, but sending qubits around through a quantum channel, rather than a classical one, effectively means leveraging the behavior of particles when taken at their smallest scale – so-called "quantum states".

Unsurprisingly, qubits cannot be used to send the kind of data we are familiar with, like emails and WhatsApp messages. But the strange behavior of qubits is opening up huge opportunities in other, more niche applications. [13]

3.2 Quantum Communications

One of the most exciting avenues that researchers, armed with qubits, are exploring, is communications *security*. [13] Quantum security leads us to the concept of *quantum cryptography* which uses physics to develop a cryptosystem completely secure against being compromised without knowledge of the sender or the receiver of the messages.

Essentially, quantum cryptography is based on the usage of individual particles/waves of light (photon) and their intrinsic quantum properties to develop an unbreakable cryptosystem (because it is impossible to measure the quantum state of any system without disturbing that system). [16] Quantum cryptography uses photons to transmit a key. Once the key is transmitted, coding and encoding using the normal secret-key method can take place. But how does a photon become a key? How do you attach information to a photon's spin? [16] This is where binary code comes into play. Each type of a photon's spin represents one piece of information – usually a 1 or a 0, for binary code.

This code uses strings of 1s and 0s to create a coherent message. For example, 11100100110 could correspond to h-e-l-l-o. So a binary code can be assigned to each photon, for example, a photon that has a vertical spin (|) can be assigned a 1.

Regular, non-quantum encryption can work in a variety of ways but generally a message is scrambled and can only be unscrambled using a secret key. The trick is to make sure that whomever you're trying to hide your communication from doesn't get their hands on your secret key. However, such encryption techniques have their vulnerabilities. Certain products – called weak keys – happen to be easier to factor than others. Also, Moore's law continually ups the processing power of our computers. Even more importantly, mathematicians are constantly developing new algorithms that allow for easier factorization of the secret key. [16]

Quantum cryptography avoids all these issues. Here, the key is encrypted into a series of photons that get passed between two parties trying to share secret information. The Heisenberg uncertainty principle dictates that an adversary can't look at these photons without changing or destroying them. [16]

4

Quantum Teleportation

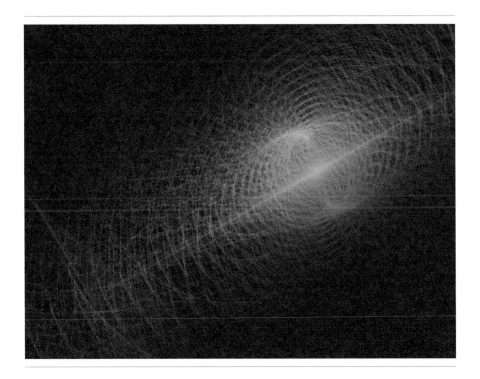

Quantum teleportation is a technique for transferring quantum information from a sender at one location to a receiver some distance away. While teleportation is commonly portrayed in science fiction as a means to transfer

Spooky action at a distance.

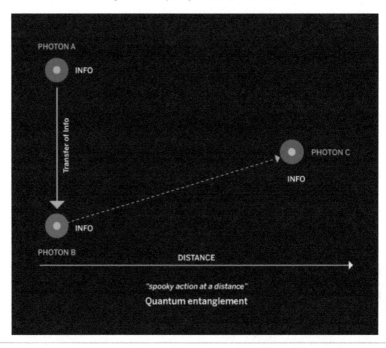

physical objects from one location to the next, quantum teleportation only transfers *quantum information.* An interesting note is that the sender knows neither the location of the recipient nor the quantum state that will be transferred. [17]

For the first time, a team of scientists and researchers have achieved sustained, high-fidelity "quantum teleportation" – the instant transfer of "qubits", the basic unit of quantum information. The collaborative team, which includes NASA's jet propulsion laboratory, successfully demonstrated sustained, long-distance teleportation of qubits of photons (quanta of light) with fidelity greater than 90%. The qubits (quantum bits) were teleported 44 km (27 miles) over a fiber-optic network using state-of-the-art single-photon detectors and off-the-shelf equipment. [20]

Quantum teleportation is the transfer of quantum states from one location to another. Through quantum entanglement, two particles in separate locations are connected by an invisible force, famously referred to as "spooky action at a distance" by Albert Einstein (Figure 4.1). Regardless of the distance, the encoded information shared by the "entangled" pair of particles can be passed between them. [20]

By sharing these quantum qubits, the basic units of quantum computing, researchers are hoping to create networks of quantum computers that can share information at blazing-fast speeds. However, keeping this information flow stable over long distances has proven extremely difficult, and researchers are now hoping to scale up such a system, using both entanglement to send information and quantum memory to store it as well. [20]

On the same front, researchers have advanced their quantum technology researches with a chip that could be scaled up and used to build the quantum simulator of the future using nanochip that allows them to produce enough stable photons encoded with quantum information to scale up the technology. The chip, which is said to be less than one-tenth of the thickness of a human hair, may enable the researchers to achieve "quantum supremacy" – where a quantum device can solve a given computational task faster than the world's most powerful supercomputer. [20]

But first let's explain some of the basic terms of the quantum world: *Quantum computing* is the area of study focused on developing computer technology based on the principles of *quantum theory*. The quantum computer, following the laws of quantum physics, would gain *enormous processing power* through the ability to be in multiple states, and to perform tasks using all possible permutations simultaneously. [18]

4.1 Quantum Teleportation: Paving the Way for a Quantum Internet

In July, the US Department of Energy unveiled a blueprint for the first quantum internet, connecting several of its National Laboratories across the country. A quantum internet would be able to transmit large volumes of data across immense distances at a rate that exceeds the *speed of light*. You can imagine all the applications that can benefit from such speed. [18]

Traditional computer data is coded in either zeros or ones. Quantum information is superimposed in both zeros and ones simultaneously. Academics, researchers and IT professionals will need to create devices for the infrastructure of quantum internet including: quantum routers, repeaters, gateways, hubs, and other quantum tools. A whole new industry will be born based on the idea of the quantum internet existing in parallel to the current ecosystem of companies we have in regular internet. [18] The "traditional internet", as the regular internet is sometimes called, will still exist. It is expected that large organizations will rely on the quantum internet to safeguard data, but that individual consumers will continue to use the classical internet. [18]

Experts predict that the financial sector will benefit from the quantum internet when it comes to securing online transactions. The healthcare sectors and the public sectors are also expected to see benefits. In addition to providing a faster, safer internet experience, quantum computing will better position organizations to solve complex problems like supply chain management. Furthermore, it will expedite the exchange of vast amounts of data, and carrying out large-scale sensing experiments in astronomy, materials discovery and life sciences. [18]

5

Quantum Computing and the IoT

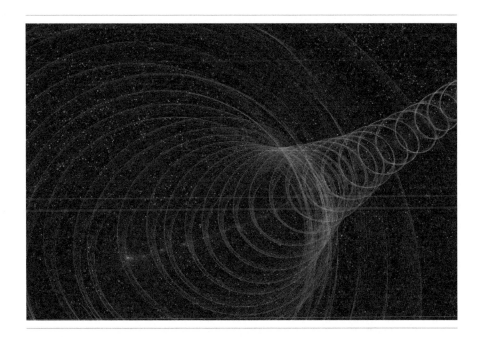

Consumers, companies, and governments will install 40 billion IoT devices globally. Smart tech finds its way into every business and consumer domain there is – from retail to healthcare, from finances to logistics – and a missed opportunity strategically employed by a competitor can easily qualify as a long-term failure for companies who don't innovate.

Moreover, the 2020 challenges, including : data breaches, malware and ransomware that paralyzed IoT components namely sensors, networks, cloud, and applications, just confirmed the need to secure all four components of the IoT: model: sensors, networks (communications), analytics (cloud), and applications [21, 23].

One of the top candidates to help in securing the IoT is quantum computing; while the idea of convergence of IoT and quantum computing is not a new topic, it has been discussed in many works of literature and covered by various researchers, but nothing is close to a practical application so far. Quantum computing is not ready yet, it is years away from deployment on a commercial scale. To understand the complexity of this kind of convergence, first you need to recognize the security issues of IoT, and second comprehend the complicated nature of quantum computing.

The IoT system's diverse security issues include [4, 5, 7]:

- **Data breaches:** IoT applications collect a lot of user data, most of it sensitive or personal, to operate and function correctly. As such, it needs encryption protection.
- **Data authentication:** Some devices may have adequate encryption in place but it can still be open to hackers if the authenticity of the data that is communicated to and from the IoT device cannot be authenticated.
- **Side-channel attacks:** Certain attacks focus on the data and information it can gain from a system's implementation rather than vulnerabilities in the implementation's algorithms.
- **Irregular updates:** Due to rapid advances in the IoT industry, a device that may have been secure on its release may no longer be secure if its software is not updated regularly. Add to that the famous SolarWinds supply chain attack of 2020 which infected over 18,000 companies and government agencies using updates of office applications and network monitoring tools.
- **Malware and ransomware:** Malware refers to the multitude of malicious programs that typically infect a device and influence its functioning whereas ransomware has the capability to lock a user out of their device, usually requesting a "ransom" to gain full use back again, paid by the cryptocurrency "Bitcoin".

5.1 A Comparison of Classical and Quantum Computing

Classical computing relies, at its ultimate level, on principles expressed by a branch of math called Boolean algebra. Data must be processed in an exclusive binary state at any point in time or bits. While the time that each transistor or capacitor needs to be either in 0 or 1 before switching states is now measurable in billionths of a second, there is still a limit as to how quickly these devices can be made to switch state. As we progress to smaller and faster circuits, we begin to reach the physical limits of materials and the threshold for classical laws of physics to apply. Beyond this, the quantum world takes over.

In a quantum computer, several elemental particles such as electrons or photons can be used with either their charge or polarization acting as a representation of 0 and/or 1. Each of these particles is known as a quantum bit, or qubit, the nature and behavior of these particles form the basis of quantum computing [22].

5.2 Quantum Communications

One of the most exciting avenues that researchers, armed with qubits, are exploring is communications security. Quantum security leads us to the concept of *quantum cryptography* which uses physics to develop a cryptosystem completely secure against being compromised without the knowledge of the sender or the receiver of the messages. Essentially, quantum cryptography is based on the usage of individual particles/waves of light (photon) and their intrinsic quantum properties to develop an unbreakable cryptosystem (because it is impossible to measure the quantum state of any system without disturbing that system).

Quantum cryptography uses photons to transmit a key. Once the key is transmitted, coding and encoding using the normal secret-key method can take place. But how does a photon become a key? How do you attach information to a photon's spin? This is where binary code comes into play. Each type of a photon's spin represents one piece of information – usually a 1 or 0, for binary code. This code uses strings of 1s and 0s to create a coherent message. For example, 11100100110 could correspond to h-e-l-l-o. So a binary code can be assigned to each photon – for example, a photon that has a vertical spin (|) can be assigned a 1.

Quantum cryptography avoids all these issues. Here, the key is encrypted into a series of photons that get passed between two parties trying to share secret information. Heisenberg's uncertainty principle dictates that an adversary can't look at these photons without changing or destroying them [22, 24].

Quantum Computing and IoT

With its capabilities, quantum computing can help address the challenges and issues that hamper the growth of IoT. Some of these capabilities are [3]:

- **Optimized complex computation power:** With quantum computing the speed is incredibly high; IoT benefits from this speed since IoT devices generate a massive amount of data that requires heavy computation and other complex optimization.
- **Faster validation and verification process:** Quantum computing addresses this concern as it can speed up the verification and validation process across all the systems several times faster while ensuring constant optimization of the systems.
- **More secure communications:** A more secure communication is possible through quantum cryptography, as explained before. The complexity serves as a defense against cyberattacks including data breaches, authentication, and malware, and ransomware.

The Road Ahead

Quantum computing is still in its development stage with tech giants such as IBM, Google, and Microsoft putting in resources to build powerful quantum computers. While they have been able to build machines containing more and more qubits, for example, Google announced in 2019 they achieved "quantum supremacy", the challenge is to get these qubits to operate smoothly and with fewer errors. But with the technology being very promising, continuous research and development are expected until such time that it reaches widespread practical applications for both consumers and businesses [23, 26].

IoT is expanding as we depend on our digital devices more every day. Furthermore, the WFH (work from home) concept resulting from COVID-19 lockdowns accelerated the deployment of many IoT devices and shortened the learning curves of using such devices. When IoT converges with quantum computing under "quantum IoT" or QIoT, this will push other technologies to use quantum computing and add "quantum" or "Q" to their products and services labels. We will then see more adoption of quantum hardware and software applications in addition to quantum services like QSaaS, QIaaS, and QPaaS as parts of quantum cloud and QAI (quantum artificial intelligence) to mention few examples. [25 , 27]

6

Quantum Computing and Blockchain: Myths and Facts

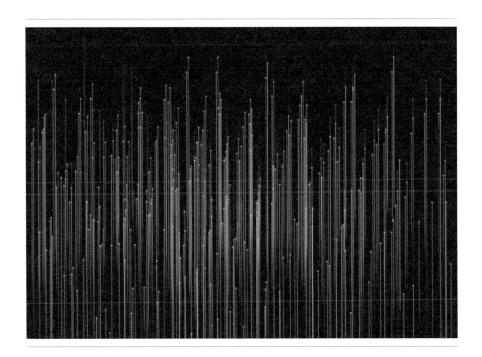

The biggest danger to Blockchain networks from quantum computing is its ability to break traditional encryption [30].

Google sent shock waves around the internet when it was claimed they had built a quantum computer able to solve formerly impossible mathematical calculations – with some fearing the crypto industry could be at risk. Google stated that its experiment was the first experimental challenge against the *extended Church–Turing thesis* – also known as the computability thesis – which claims that traditional computers can effectively carry out any "reasonable" model of computation.

6.1 Difficulties with Quantum Computers [31]

- **Interference**: During the computation phase of a quantum calculation, the slightest disturbance in a quantum system (say a stray photon or wave of EM radiation) causes the quantum computation to collapse, a process known as de-coherence. A quantum computer must be totally isolated from all external interference during the computation phase.
- **Error correction**: Given the nature of quantum computing, error correction is ultra-critical – even a single error in a calculation can cause the validity of the entire computation to collapse.
- **Output observance**: Closely related to the above two, retrieving output data after a quantum calculation is complete risks corrupting the data.

6.2 What is Quantum Supremacy?

According to the *Financial Times*, Google claims to have successfully built the world's most powerful quantum computer. What that means, according to Google's researchers, is that calculations that normally take more than 10,000 years to perform its computer was able to do in about *200 s*, and this potentially means Blockchain, and the encryption that underpins it, could be broken.

Asymmetric cryptography used in crypto relies on key pairs, namely a private and public key. Public keys can be calculated from their private counterpart, but *not* the other way around. This is due to the impossibility of certain mathematical problems. Quantum computers are more efficient in accomplishing this by magnitudes, and if the calculation is done the other way then the whole scheme breaks [30].

It would appear Google is still some way away from building a quantum computer that could be a threat to Blockchain cryptography or other encryption.

"Google's supercomputer currently has 53 qubits," said Dragos Ilie, a quantum computing and encryption researcher at Imperial College London.

"In order to have any effect on bitcoin or most other financial systems it would take at least about 1500 qubits and the system must allow for the entanglement of all of them," Ilie said.

Meanwhile, scaling quantum computers is "a huge challenge," according to Ilie [28].

Blockchain networks, including Bitcoin's architecture, relies on two algorithms: the elliptic curve digital signature algorithm (ECDSA) for digital signatures and SHA-256 as a hash function. A quantum computer could use Shor's algorithm [8] to get your private information from your public key, but the most optimistic scientific estimates say that even if this were possible, it won't happen during this decade.

"A 160 bit elliptic curve cryptographic key could be broken on a quantum computer using around *1000 qubits* while factoring the security-wise equivalent 1024 bit RSA modulus would require about *2000 qubits*". By comparison, Google's measly 53 qubits are still no match for this kind of cryptography, according to a research paper on the matter published by Cornell University.

But that isn't to say that there's no cause for alarm. While the native encryption algorithms used by Blockchain's applications are safe for now, the fact is that the rate of advancement in quantum technology is increasing, and that could, in time, pose a threat. "We expect their computational power will continue to grow at a double exponential rate," say Google researchers.

6.3 Quantum Cryptography?

Quantum cryptography uses physics to develop a cryptosystem completely secure against being compromised without knowledge of the sender or the receiver of the messages. The word *quantum* itself refers to the most fundamental behavior of the smallest particles of matter and energy.

Quantum cryptography is different from traditional cryptographic systems in that it relies more on *physics*, rather than mathematics, as a key aspect of its security model.

Essentially, quantum cryptography is based on the usage of individual particles/waves of light (photon) and their intrinsic quantum properties to develop an unbreakable cryptosystem (*because it is impossible to measure the quantum state of any system without disturbing that system*).

Quantum cryptography uses photons to transmit a key. Once the key is transmitted, coding and encoding using the normal secret-key method can take place. But how does a photon become a key? How do you attach information to a photon's spin?

This is where *binary code* comes into play. Each type of a photon's spin represents one piece of information – usually a 1 or a 0, for binary code. This code uses strings of 1s and 0s to create a coherent message. For example, 11100100110 could correspond to h-e-l-l-o. So a binary code can be assigned to each photon – for example, a photon that has a *vertical spin* (|) can be assigned a 1.

"If you build it correctly, no hacker can hack the system. The question is what it means to build it correctly," said physicist Renato Renner from the Institute of Theoretical Physics in Zurich.

Regular, non-quantum encryption can work in a variety of ways but generally a message is scrambled and can only be unscrambled using a secret key. The trick is to make sure that whomever you're trying to hide your communication from doesn't get their hands on your secret key. Cracking the private key in a modern crypto system would generally require figuring out the factors of a number that is the product of two insanely huge prime numbers.

The numbers are chosen to be so large that, with the given processing power of computers, it would take longer than the lifetime of the universe for an algorithm to factor their product.

Encryption techniques have their vulnerabilities. Certain products – called weak keys – happen to be easier to factor than others. Also, Moore's Law continually ups the processing power of our computers. Even more importantly, mathematicians are constantly developing new algorithms that allow for easier factorization.

Quantum cryptography avoids all these issues. Here, the key is encrypted into a series of photons that get passed between two parties trying to share secret information. The Heisenberg uncertainty principle dictates that an adversary can't look at these photons without changing or destroying them.

"In this case, it doesn't matter what technology the adversary has, they'll never be able to break the laws of physics," said physicist Richard Hughes of Los Alamos National Laboratory in New Mexico, who works on quantum cryptography [32, 33].

7

Quantum Computing and AI: A Mega-Buzzword

Quantum computers are designed to perform tasks much more accurately and efficiently than conventional computers, providing developers with a new tool for specific applications. It is clear in the short-term that quantum computers will not replace their traditional counterparts; instead, they will require classical computers to support their specialized abilities, such as system optimization. [35]

Quantum computing and artificial intelligence are both transformational technologies and artificial intelligence needs quantum computing to achieve significant progress. Although artificial intelligence produces functional applications with classical computers, it is limited by the computational capabilities of classical computers. Quantum computing can provide a computation boost to artificial intelligence, enabling it to tackle more complex problems in many fields in business and science. [37]

7.1 What is Quantum Computing?

Quantum computing is the area of study focused on developing computer technology based on the principles of quantum theory. The quantum computer, following the laws of quantum physics, would gain enormous processing power through the ability to be in multiple states and to perform tasks using all possible permutations simultaneously.

7.2 Difficulties with Quantum Computers

- **Interference**: During the computation phase of a quantum calculation, the slightest disturbance in a quantum system (say a stray photon or wave of EM radiation) causes the quantum computation to collapse, a process known as de-coherence. A quantum computer must be totally isolated from all external interference during the computation phase.
- **Error correction**: Given the nature of quantum computing, error correction is ultra-critical – even a single error in a calculation can cause the validity of the entire computation to collapse.
- **Output observance**: Closely related to the above two, retrieving output data after a quantum calculation is complete risks corrupting the data.

7.3 Applications of Quantum Computing and AI

Keeping in mind that the term "quantum AI" means the use of quantum computing for computation of machine learning algorithms, which takes advantage of computational superiority of quantum computing to achieve results that are not possible to achieve with classical computers, the following are some of the applications of this super mix of quantum computing and AI (Figure 7.1) [34, 37]:

7.4 Processing Large Sets of Data

We produce 2.5 exabytes of data every day. That's equivalent to 250,000 Libraries of Congress or the content of 5 million laptops. Every minute of every day 3.2 billion global internet users continue to feed the data banks with

Figure 7.1: Applications of quantum computing and AI.

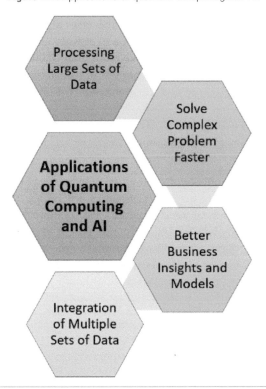

9722 pins on Pinterest, 347,222 tweets, 4.2 million Facebook likes plus ALL the other data we create by taking pictures and videos, saving documents, opening accounts and more. [36]

Quantum computers are designed to manage the huge amount of data, along with uncovering patterns and spotting anomalies extremely quickly. With each newly launched iteration of quantum computer design and the new improvements made on the quantum error-correction code, developers are now able to better manage the potential of quantum bits. They also optimize the same for solving all kinds of business problems to make better decisions. [35]

7.5 Solve Complex Problem Faster

Quantum computers can complete calculations within seconds, which would take today's computers many years to calculate. With quantum computing, developers can do multiple calculations with multiple inputs simultaneously. Quantum computers are critical to process the monumental amount of data that businesses generate on a daily basis, and the fast calculation can be used to solve very complex problems, which can be expressed as quantum supremacy; where the calculations that normally take more than 10,000 years to perform, a quantum computer can do it *200 s.* The key is to translate real-world problems that companies are facing into quantum language. [35, 39]

7.6 Better Business Insights and Models

With the increasing amount of data generated in industries like pharmaceutical, finance and life science industries, companies are losing their ties with the classical computing rope. To have a better data framework, these companies now require complex models that have the potential processing power to model the most complex situations. And that's where quantum computers play a huge role. Creating better models with quantum technology will lead to better treatments for diseases in the healthcare sector, like the COVID-19 research cycle from test, tracing and treating of the virus, can decreased financial implosion in the banking sector and improve the logistics chain in the manufacturing industry. [35]

7.7 Integration of Multiple Sets of Data

To manage and integrate multiple numbers of sets of data from multiple sources, quantum computers are best to help, as they make the process quicker, and also make the analysis easier. The ability to handle so many stakes has made quantum computing an suitable choice for solving business problems in a variety of fields. [35]

7.8 The Future

The quantum computing market will reach $2.2 billion, and the number of installed quantum computers will reach around 180, in 2026, with about 45 machines produced in that year. These include both machines installed at the quantum computer companies themselves that are accessed by quantum services as well as customer premises machines. [38]

Cloud access revenues will likely dominate as a revenue source for quantum computing companies in the format of quantum computing as a service (QCaaS) offering, which will account for 75% of all quantum computing revenues in 2026. Although, in the long run, quantum computers may be more widely purchased, today potential end users are more inclined to do quantum computing over the cloud rather than make technologically risky and expensive investments in quantum computing equipment. [38]

On a parallel track, quantum software applications, developers' tools and the number of quantum engineers and experts will grow as the infrastructure develops over the next 5 years, which will make it possible for more organizations to harvest the power of two transformational technologies, quantum computing and AI, and encourage many universities to add quantum computing as an essential part of their curriculum.

8

Quantum Computing Trends

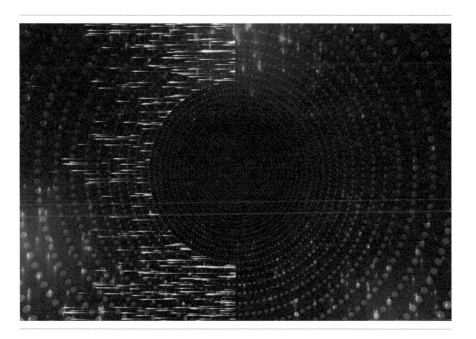

Quantum computing is the area of study focused on developing computer technology based on the principles of quantum theory. Tens of billions of public and private capitals are being invested in quantum technologies. Countries across the world have realized that quantum technologies can be a major disruptor of existing businesses, and they collectively invested $24 billion in in quantum research and applications in 2021 [40].

Figure 8.1: Future of computing.

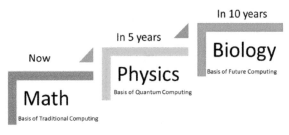

Future of Computing

8.1 A Comparison of Classical and Quantum Computing

Classical computing relies, at its ultimate level, on principles expressed by Boolean algebra (Figure 8.1). Data must be processed in an exclusive binary state at any point in time or what we call bits. While the time that each transistor or capacitor need be either in 0 or 1 before switching states is now measurable in billionths of a second, there is still a limit as to how quickly these devices can be made to switch state.

As we progress to smaller and faster circuits, we begin to reach the physical limits of materials and the threshold for classical laws of physics to apply. Beyond this, the quantum world takes over; in a quantum computer, a number of elemental particles such as electrons or photons can be used with either their *charge* or *polarization* acting as a representation of 0 and/or 1. Each of these particles is known as a quantum bit, or *qubit*, the nature and behavior of these particles form the basis of quantum computing [2]. Classical computers use transistors as the physical building blocks of logic, while quantum computers may use trapped ions, superconducting loops, quantum dots or vacancies in a diamond [40].

8.2 Physical vs. Logical Qubits

When discussing quantum computers with error correction, we talk about physical and logical qubits. Physical qubits are the physical qubits in a quantum computer, whereas logical qubits are groups of physical qubits we use as a single qubit in our computation to fight noise and improve error correction.

To illustrate this, let's consider an example of a quantum computer with 100 qubits. Let's say this computer is prone to noise. To remedy this we can use multiple qubits to form a single more stable qubit. We might decide that we need 10 physical qubits to form one acceptable logical qubit. In this case we would say our quantum computer has 100 physical qubits which we use as 10 logical qubits.

Distinguishing between physical and logical qubits is important. There are many estimates as to how many qubits we will need to perform certain calculations, but some of these estimates talk about logical qubits and others talk about physical qubits. For example, to break RSA cryptography we would need thousands of logical qubits but millions of physical qubits.

Another thing to keep in mind, in a classical computer computeation power increases linearly with the number of transistors and clock speed, while in a quantum computer computational power increases exponentially with the addition of each logical qubit [43].

8.3 Quantum Superposition and Entanglement

The two most relevant aspects of quantum physics are the principles of *superposition* and *entanglement*.

Superposition: Think of a qubit as an electron in a magnetic field. The electron's spin may be either in alignment with the field, which is known as a spin-up state, or opposite to the field, which is known as a spin-down state. According to quantum law, the particle enters a superposition of states, in which it behaves as if it were in both states simultaneously. Each qubit utilized could take a superposition of both 0 and 1. Where a 2-bit register in an ordinary computer can store only one of four binary configurations (00, 01, 10, or 11) at any given time, a 2-qubit register in a quantum computer can store all four numbers simultaneously, because each qubit represents two values. If more qubits are added, the increased capacity is expanded exponentially.

Entanglement: Particles that have interacted at some point retain a type of connection and can be entangled with each other in pairs, in a process known as correlation. Knowing the spin state of one entangled particle – up or down – allows one to know that the spin of its mate is in the opposite direction. Quantum entanglement allows qubits that are separated by incredible distances to interact with each other instantaneously (not limited to the speed of light). No matter how great the distance between the correlated particles, they will remain entangled as long as they are isolated. Taken together, quantum

Figure 8.2: Quantum computer categories.

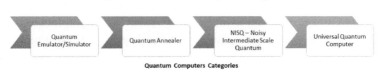

Quantum Computers Categories

superposition and entanglement create an enormously enhanced computing power [42].

Quantum computers fall into four categories [40] (Figure 8.2):

1. Quantum emulator/simulator
2. Quantum annealer
3. Noisy intermediate scale quantum (NISQ)
4. Universal quantum computer – which can be a cryptographically relevant quantum computer (CRQC).

8.4 Quantum Emulator/Simulator

These are classical computers that you can buy today that simulate quantum algorithms. They make it easy to test and debug a quantum algorithm that someday may be able to run on a universal quantum computer (UQC). Since they don't use any quantum hardware, they are no faster than standard computers.

8.5 Quantum Annealer

A special purpose quantum computer designed to only run combinatorial optimization problems, not general-purpose computing, or cryptography problems. While they have more physical qubits than any other current system they are not organized as gate-based logical qubits. Currently this is a commercial technology in search of a future viable market.

8.6 Noisy Intermediate-scale Quantum (NISQ) Computers

Think of these as *prototypes* of a universal quantum computer – with several orders of magnitude fewer bits. They currently have 50–100 qubits, limited gate depths, and short coherence times. As they are short several orders of magnitude of qubits, NISQ computers cannot perform any useful computation; however they are a necessary phase in the learning, especially to drive total system and software learning in parallel to the hardware development. Think of them as the training wheels for future universal quantum computers.

8.7 Universal Quantum Computers/Cryptographically Relevant Quantum Computers (CRQC)

This is the ultimate goal. If you could build a universal quantum computer with fault tolerance (i.e., millions of error-corrected physical qubits resulting in thousands of logical qubits), you could run quantum algorithms in cryptography, search and optimization, quantum systems simulations, and linear equations solvers.

8.8 Post-quantum/Quantum-resistant Codes

New cryptographic systems would secure against both quantum and conventional computers and can interoperate with existing communication protocols and networks. The symmetric key algorithms of the Commercial National Security Algorithm (CNSA) Suite were selected to be secure for national security systems usage even if a CRQC is developed. Cryptographic schemes that commercial industry believes are quantum-safe include lattice-based cryptography, hash trees, multivariate equations, and super-singular isogeny elliptic curves [40].

8.9 Difficulties with Quantum Computers [41]

- **Interference**: During the computation phase of a quantum calculation, the slightest disturbance in a quantum system (say a stray photon or wave of EM radiation) causes the quantum computation to collapse, a process known as de-coherence. A quantum computer must be totally isolated from all external interference during the computation phase.
- **Error correction**: Given the nature of quantum computing, error correction is ultra-critical – even a single error in a calculation can cause the validity of the entire computation to collapse.
- **Output observance**: Closely related to the above two, retrieving output data after a quantum calculation is complete risks corrupting the data.

References

[1]   http://www.fastcolabs.com/3013214/why-quantum-computing-is-faster-for-everything-but-the-web

[2]   http://www.theguardian.com/science/2014/mar/06/quantum-computing-explained-particle-mechanics

[3]   http://www.economist.com/news/science-and-technology/21578027-first-real-world-contests-between-quantum-computers-and-standard-ones-faster

[4]   http://whatis.techtarget.com/definition/quantum-computing

[5]   http://physics.about.com/od/quantumphysics/f/quantumcomp.htm

[6]   http://www.qi.damtp.cam.ac.uk/node/38

[7]   http://www.businessinsider.com/what-is-quantum-encryption-2014-3#ixzz33jYuMw48

[8]   http://www.wired.com/2013/06/quantum-cryptography-hack/

[9]   http://searchsecurity.techtarget.com/definition/quantum-cryptography

[10]   http://science.howstuffworks.com/science-vs-myth/everyday-myths/quantum-cryptology.htm

[11]   http://www.wisegeek.com/what-is-quantum-cryptography.htm

[12]   http://www.techrepublic.com/blog/it-security/how-quantum-cryptography-works-and-by-the-way-its-breakable/

[13]   https://www.cybertalk.org/2020/10/23/quantum-internet-fast-forward-into-the-future/

[14]   https://www.bbvaopenmind.com/en/technology/digital-world/quantum-computing/

[15]   https://www.zdnet.com/article/what-is-the-quantum-internet-everything-you-need-to-know-about-the-weird-future-of-quantum-networks/

[16]   https://ahmedbanafa.blogspot.com/2014/06/understanding-quantum-cryptography.html

[17]   https://en.wikipedia.org/wiki/Quantum_teleportation

[18]   https://www.linkedin.com/pulse/quantum-internet-explained-ahmed-banafa/

[19]   https://www.designboom.com/technology/nasa-long-distance-quantum-teleportation-12-22-2020/

[20]   https://www.siliconrepublic.com/machines/quantum-computing-fermilab

[21]   https://ahmedbanafa.blogspot.com/2019/12/ten-trends-of-iot-in-2020.html

[22]   https://ahmedbanafa.blogspot.com/2020/11/quantum-internet-explained.html

[23]   https://www.azoquantum.com/Article.aspx?ArticleID=101
[24]   https://www.cybersecurityintelligence.com/blog/quantum-computing-the-internet-of-things-and-hackers-4914.html
[25]   https://www.europeanbusinessreview.com/iot-security-are-we-ready-for-a-quantum-world/
[26]   https://www.bbvaopenmind.com/en/technology/digital-world/quantum-computing-and-blockchain-facts-and-myths/
[27]   https://www.ft.com/content/c13dbb51-907b-4db7-8347-30921ef
[28]   https://www.forbes.com/sites/billybambrough/2019/10/02/could-google-be-about-to-break-bitcoin/#1d78c5373329
[29]   https://decrypt.co/9642/what-google-quantum-computer-means-for-bitcoin/
[30]   https://www.coindesk.com/how-should-crypto-prepare-for-googles-quantum-supremacy?
[31]   https://www.ccn.com/google-quantum-bitcoin/
[32]   https://www.linkedin.com/pulse/20140503185010-246665791-quantum-computing/
[33]   https://www.linkedin.com/pulse/20140608053056-246665791-understanding-quantum-
[34]   https://www.linkedin.com/pulse/quantum-computing-blockchain-facts-myths-ahmed-banafa/
[35]   https://analyticsindiamag.com/will-quantum-computing-define-the-future-of-ai/
[36]   https://www.analyticsinsight.net/ai-quantum-computing-can-enable-much-anticipated-advancements/
[37]   https://research.aimultiple.com/quantum-ai/
[38]   https://www.globenewswire.com/news-release/2020/11/17/2128495/0/en/Quantum-Computing-Market-is-Expected-to-Reach-2-2-Billion-by-2026.html
[39]   https://ai.googleblog.com/2019/10/quantum-supremacy-using-programmable.html
[40]   https://www.linkedin.com/pulse/quantum-technology-ecosystem-explained-steve-blank/?
[41]   https://www.bbvaopenmind.com/en/technology/digital-world/quantum-computing-and-ai/
[42]   https://phys.org/news/2022-03-technique-quantum-resilient-noise-boosts.html
[43]   https://thequantuminsider.com/2019/10/01/introduction-to-qubits-part-1/

Index